Reefs in Europe:

None

Reefs in Asia:

4. Persian Gulf
5. Sri Lanka
6. Maldives
7. Thailand
8. Malaysia
9. Philippines
10. Indonesia

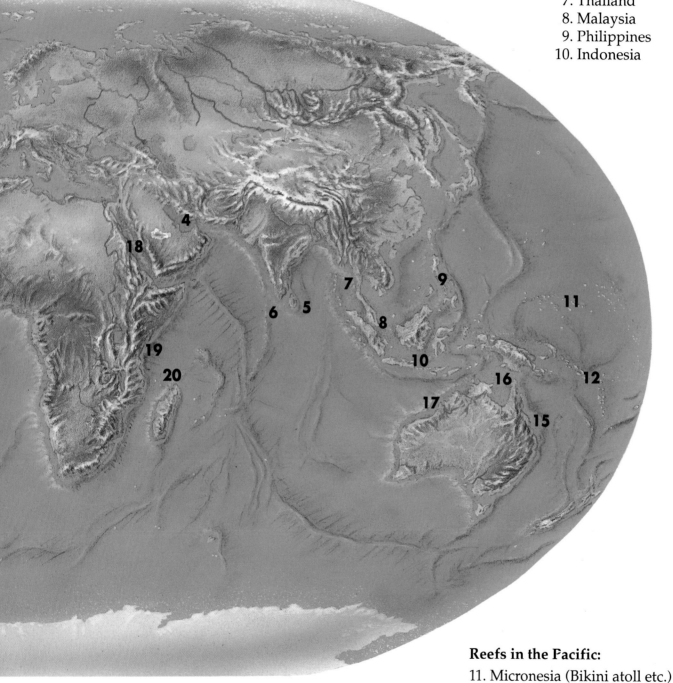

Reefs in the Pacific:

11. Micronesia (Bikini atoll etc.)
12. Melanesia (Soloman islands, Fiji etc.)
13. Polynesia (Tonga, Western Samoa, Bora Bora, Tahiti etc.)
14. Hawaiian Islands

Reefs in Australasia:

15. Great Barrier Reefs (Queensland)
16. Northern Territory
17. Western Australia

3

FACTS ABOUT REEFS

Coral reefs are most commonly found in the tropics where the water is clear, warm and with plenty of strong sunlight. Corals grow most strongly of all where the water temperature lies between 73 and 90 °F and in waters no deeper than 60 ft. Corals only thrive in salty seas and they will not grow where rainfall is high (making the sea surface waters less salty) or where rivers bring fresh water into the sea.

Corals rely on strong wave action and ocean currents to clean them of any silt that would otherwise settle out of the water and to bring them a constant supply of tiny food particles.

The western sides of ocean basins have the most extensive coral reefs. The world's largest continuous coral reef area is the Australian Great Barrier Reefs. This group stretches for 1200 mi parallel to the north-east shore of Queensland, on the western side of the Pacific Ocean.

A reef is one of the most complex communities of plants and animals in the world. The color of the reefs comes not from the corals, but from algae that live in the tissues of the coral. Algae supply the coral with oxygen and nutrients in a unique relationship.

Corals have built huge reefs in the past and many of them have been preserved as limestone rock.

Grolier Educational Corporation
SHERMAN TURNPIKE, DANBURY, CONNECTICUT 06816

LAND ⊕ SHAPES

REEF

Author
Brian Knapp, BSc, PhD
Art Director
Duncan McCrae, BSc
Editor
Rita Owen
Illustrators
David Hardy and David Woodroffe
Print consultants
Landmark Production Consultants Ltd
Printed and bound in Hong Kong
Designed and produced by
EARTHSCAPE EDITIONS

First published in the USA in 1993 by
GROLIER EDUCATIONAL CORPORATION,
Sherman Turnpike, Danbury, CT 06816

Copyright © 1992
Atlantic Europe Publishing Company Limited

Library of Congress #92–072045

Cataloging information may be obtained
directly from Grolier Educational Corporation

Title ISBN   0–7172–7186–2

Set ISBN   0–7172–7176–5

Acknowledgements. The publishers would like
to thank Redlands County Primary School.

Picture credits. All photographs from the
Earthscape Editions photographic library except
the following (t=top, b=bottom, l=left, r=right):
Jack Jackson *inside back cover*, 5, 6, 8/9, 12/13, 14/15,
14b, 15t, 15b, 17, 18/19, 32/33, 32b; NASA 31t;
ZEFA *Cover*, 10/11, 16, 18b, 20, 21, 22, 24/25,
26/27, 26b, 27r, 28, 29t, 30/31, 31b.

Cover picture: Great Barrier Reefs,
Queensland, Australia.
Inside back cover: Polyps in the Red Sea, Sudan.

In this book you will find some
words that have been shown in **bold**
type. There is a full explanation of
each of these words on page 36.

On some pages you will
find experiments that you
might like to try for
yourself. They have been
put in a blue box like this.

In this book mi means miles and
ft means feet.

Divers appear on a number of
pages to help you to know the size
of some landshapes.

CONTENTS

Chapter 1: How reefs form

Chapter 2: Types of reef

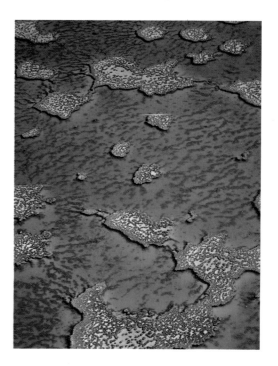

Chapter 3: Reefs of the world

Introduction

Reef comes from the old Viking word *rif* which means a rib – an undersea danger to ships. When European sailors first arrived at shores surrounded by coral reefs they were very aware that the sharp corals could tear the bottoms of their fragile boats. In time the word rif became changed to reef.

Coral reefs are frames made of countless millions of sea animals and plants that make limestone skeletons. By each new generation fastening itself to the remains of its dead predecessor's skeletons, coral reefs form massive structures that have made some of the most important rocks through geological time.

Corals, and their companion plants called algae, are some of the world's most incredible living things. These minuscule creatures manage to build reefs which can survive even hurricane-sized waves, a feat that many man-made structures cannot match.

Corals also combine strength with beauty to make colorful branching shapes. Together they make the world's only living landshape. Just turn to a page to enjoy the wonder of coral reefs.

Take care by reefs

It is fun to visit coral reefs while on vacation to see the landshapes described in this book for yourself. But never venture on to a coral reef without an adult and never go snorkelling without a qualified guide. The waves can be fierce and they can throw you on to the sharp coral, while the currents can drag you out to sea. Remember, too, that the reef is living and some animals defend themselves with poisons when touched.

Chapter 1:
How reefs form

The landshapes of reefs

Corals and algae are living things. They need special conditions, such as warm clear water and light to thrive. Coral reefs are the frameworks that develop when huge numbers of these creatures find just the right conditions to grow.

Any rocky base, or even a mound of rubble, can be the home of reef-building coral provided it is in warm, shallow clear water. So coral creatures are just as at home on island mountains as they are on the shallow margins of continents.

Crest of reef which is a harsh environment for corals (see page 18).

Gaps in the reef allow water in and out of the lagoon (see page 21).

The main area of coral is called a barrier reef if there is a large lagoon behind it (page 26), an atoll if there is no area of land inside the reef (page 28), or a fringing reef if there is no lagoon, or only a small lagoon between the reef and the shore (see page 26).

Reef history
The reefs that are built on coral islands in the middle of the oceans have a very different history to those near the continents. You will find them described on pages 24 to 28.

Reefs shapes

The land on which the reef builds has a big influence on the shape of the final reef. For example, the biggest reef area of all – the Great Barrier Reefs – has been able to grow so large because of the wide **continental shelf** that stretches under the sea off the Australian coast. By contrast, reef-building corals on small islands in the middle of deep oceans cannot spread out very far; they can make no more than a ring of coral.

Land that has steep sides allows many reefs to flourish (see page 24).

The reef front (see page 16).

This picture shows reefs surrounding an island. There are some differences when reefs form on shelves near to large areas of land.

Corals

Of all the world's most important landshapes, coral reefs are unique because they are built by very simple animals and plants.

To know how reefs grow and change it is therefore very important to understand a little about the way corals live.

Coral skeletons

The skeleton of the coral is both an anchor when the polyp is waving about in the water, and also a hiding place into which it can retreat when threatened. Because corals live in huge colonies, young corals build their skeleton homes on the old skeletons of their ancestors and in this way they can build a reef into a huge mass of limestone.

There are two basic types of coral – the hard, or stony, and soft corals. The hard corals produce a limestone skeleton and are the reef builders. The soft corals may look much the same but they do not have a solid limestone skeleton, and only contain tiny limestone crystals within their tissues.

Be kind to corals

Reefs are easily damaged by collectors. Never collect samples. The specimens photographed for this book come from scientific collections. None were collected just to keep at home.

What are corals?

Corals are composed of small animals called **polyps**, usually less than ⅛ inch in diameter, that belong to the same group as jellyfish and sea anemones.

In this picture the corals look dead because the tentacles of the animals have retreated within their limestone shells. But each polyp has a ring of tentacles which surrounds a central 'mouth' and this is shown on the inside rear cover. Each tentacle is fringed with tiny stinging hairs. Their job is to catch tiny pieces of food, such as **plankton**, and push them to the mouth.

13

The coral reef community

Corals are just one part of a reef community, or **ecosystem**. Within the tissues of the hard coral are millions of single-celled plants called algae. Because they are plants they must receive sunlight to grow, and this is why reefs can only thrive where the water above them is shallow.

Many reef animals depend on corals and algae for their food. By feeding on the corals and algae, they, too help to shape the growing reef.

A union of plants and animals
Algae use the energy of the Sun to convert carbon dioxide and water into food and oxygen. In this way they provide the coral animals with a plentiful supply of oxygen. They also leak nutrients from their cells, giving the corals extra nutrients for faster growth. Algae also give the corals much of their color.

This is a butterfly fish. Its specially-shaped mouth is adapted for eating the coral polyps.

The tips of branching corals can grow by up to 4 inches a year. They form the home to numerous animals as you can see in the picture on the left.

Different species of coral (as well as some algae) often compete for space on the reef surface. Grazing fish and other animals keep the algae in check and allow the coral to survive.

Reef eaters

There are many animals on a reef that eat the corals. The most important are often sponges, that bore into the coral, but starfish (such as the Crown of Thorns starfish shown here) and parrot fish also graze over the surface.

On some reefs, the numbers of starfish can reach over 20 in each square yard of surface. At this density, starfish are almost touching! When the numbers of sponges and starfish is high, they may eat the coral faster than it can grow.

The reef front

The outer edge of the reef, or reef front, is the place where food is plentiful. But because of the ocean currents and fierce waves, it is not an easy place for corals to survive.

For more information on waves and coral sand see the book Beach *in the Landshapes set.*

Much coral is broken up by waves and cast, together with shells of other reef creatures, onto the reef top as sand.

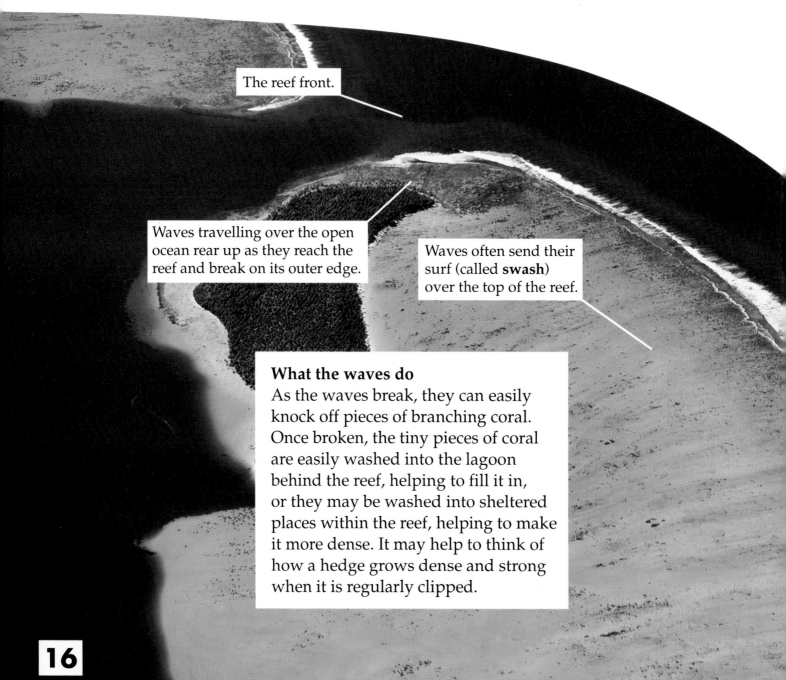

The reef front.

Waves travelling over the open ocean rear up as they reach the reef and break on its outer edge.

Waves often send their surf (called **swash**) over the top of the reef.

What the waves do
As the waves break, they can easily knock off pieces of branching coral. Once broken, the tiny pieces of coral are easily washed into the lagoon behind the reef, helping to fill it in, or they may be washed into sheltered places within the reef, helping to make it more dense. It may help to think of how a hedge grows dense and strong when it is regularly clipped.

Near the top of the slope (seen in this picture) the waves still break against the reef, making survival difficult. The variety of corals is less than in deeper water, which is why scuba diving is usually more rewarding than snorkelling.

The outer slopes are a rich source of oxygen. Also the ocean currents bring with them new supplies of nutrients. This is therefore the ideal place for corals to grow and here you find a rich variety of colors and shapes.

Away from the waves, the great Stag's Horn and other branching corals can flourish until, in deeper waters, there is not enough light for them to grow.

The importance of currents

It is not difficult to understand that breaking waves are the main force that control the growth of corals. But unseen below the water surface are powerful currents which, even at 60 ft below the surface, are as great as they are just below the waves. But far from being the places that corals avoid, these are the zones where corals grow best because currents bring their food supply.

Reef flats

Few corals can survive where waves break against the top of the reef. This is why corals stop growing upwards and spread out.

However, where there is a great tidal range each day, the top of the reef may even dry out at low tide, leaving only pools of water among the coral heads. This is a particularly harsh environment for corals, and only a small number of species occur.

Exposed reef flats

Most corals cannot survive being dried out twice a day. Therefore reef flats that become exposed at low tide are home to just one or two particularly well adapted species of coral together with the few fish that can live in the surface pools.

Thus, whereas a reef flat is the easiest place on a reef to visit, it is also the least attractive for those who are interested in the variety of reef life.

Compare this picture of the reef flat and its lack of variety with the reef front shown on page 17.

Notice how this stony coral has been partly eroded. This is probably not natural, but the destructive effect of an anchor being let out from a boat.

Flattened corals near the surface.

Cays and passes

There are many irregularities in a reef surface, with some areas much higher and others much lower. The high areas are built up with coral sand broken from the reef by the storm waves. The sand is then cemented into a hard crust by algae and a small island, or cay is formed.

Between the high areas there are often long reef-free channels known as 'passes'.

Cays

Many small islands, or cays, form where a reef is especially close to the surface.

These islands often have a crusted coral rubble on the seaward surface which protects the rest of the reef from attack. Behind this a sandy beach can develop and even enough soil to support trees, making a real 'desert island'.

Hard cemented coral rubble forms on this windward side.

The cay makes a 'desert island' by growing out of the water from a shallow reef.

A beach develops on the sheltered side.

Here you can see the passes very clearly as darker regions of water.

Passes

Reef passes show where rivers or lagoon outflows have spilled water across a reef surface.

Corals do not grow well in strong currents of fresh river water, especially if there is a large amount of silt in the water. So as the reef grows, areas that have strong currents grow less well and leave channels.

Some passes were formed during the **Ice Age**, when sea-levels were lower. The rivers ran over the surface of reefs and cut deep channels which have still not yet been filled in.

For more information on sand and wave action
see the book Beach *in the Landshapes set.*

Chapter 2:
Types of reef

The shelf reef

There are two quite different types of coral reef in the world. The world's biggest reefs are called **shelf reef**s and they form in the shallow waters that fringe most continents and *large* tropical islands. The other reefs occur dotted among the world's *small* deep-water ocean islands.

Wherever they are found, reefs occur in three patterns: fringing reefs, barrier reefs and atolls.

On a shelf reef the sea is shallow and a great variety of corals can easily grow all over the wide shelf surface. Islands have built on the reef, some as simple ovals (cays), others have become horseshoe shaped and yet others ring-shaped (atolls). This is the reason that shelf reefs contain the greatest variety of all the world's reefs.

Changes on the shelf reef

During the last Ice Age the sea level fell as ice was locked up on land near the poles. As a result, many reefs simply dried out and in turn became land. When the ice melted, sea-level rose again and corals grew on the remains of their ancestors, making new and thriving reefs.

Scientists have found that coral reefs go through cycles of growth known as the youthful, mature and old age stages, which are linked to changes in sea-level. The way this works is shown in the pictures below.

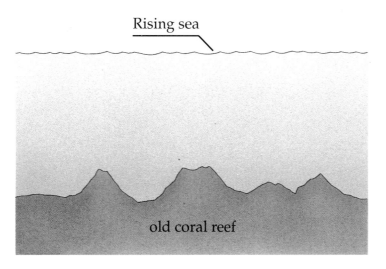

Rising sea

old coral reef

The youthful stage
Corals are quick to respond when the land sinks or the sea level rises. They soon begin growing again on top of the old coral reef. The growth can be quite rapid and usually straight upwards. This has the effect of making an already irregularly-shaped reef more and more uneven in height, as some parts of the reef grow at a faster rate than others.

The mature stage
When the reef coral grows to within about ten feet of the surface it begins to be battered by the waves and it quickly stops growing upwards and starts to spread out.

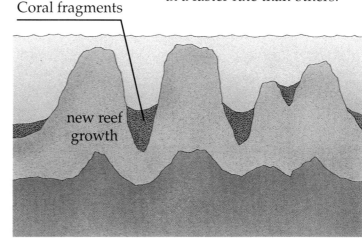

Coral fragments

new reef growth

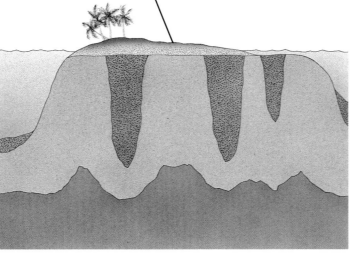

Flat top to reef.

The old age
Any coral that is broken off by the waves will fall between the main areas of growth, filling in the areas that did not grow so quickly. This makes the crest of the reef more or less flat and the cycle is complete.

The origins of island reefs

In 1835 Charles Darwin, the famous scientist, sailed through the Pacific Ocean aboard HMS *Beagle*. On his journey he found a great variety of coral islands, reefs and atolls.

Darwin already knew that the reefs he saw rose from the deep ocean floor but that corals could only grow in shallow water. How, then, could corals have formed in such places?

Sinking mountains

Darwin suggested that the pattern of reefs around ocean islands could be explained only if some of the volcanic islands were sinking. If they sank quickly, the coral would not be able to grow fast enough to keep up and no reef would form. If the coral could keep pace then it would still fringe the island to give a fringing reef. If the coral were able to grow up towards the light, but did not have enough time to spread out, it would form a ring of reef (a barrier reef) around the island. If the island finally sank out of sight, the reef would remain as a ring reef or atoll.

Not all volcanos sink at the same rate. Islands in the Pacific and Indian Oceans can therefore all be fitted into one of the schemes shown below.

1.

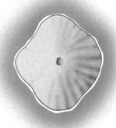

Hawaii, of the Hawaiian Islands and Tahiti, of the Society Islands are examples of volcanic islands without a reef.

This picture shows the volcanic island of Moorea with a fringing reef. (For more information on the origin of volcanos see the books Volcano *and* Mountain *in the Landshapes set.)*

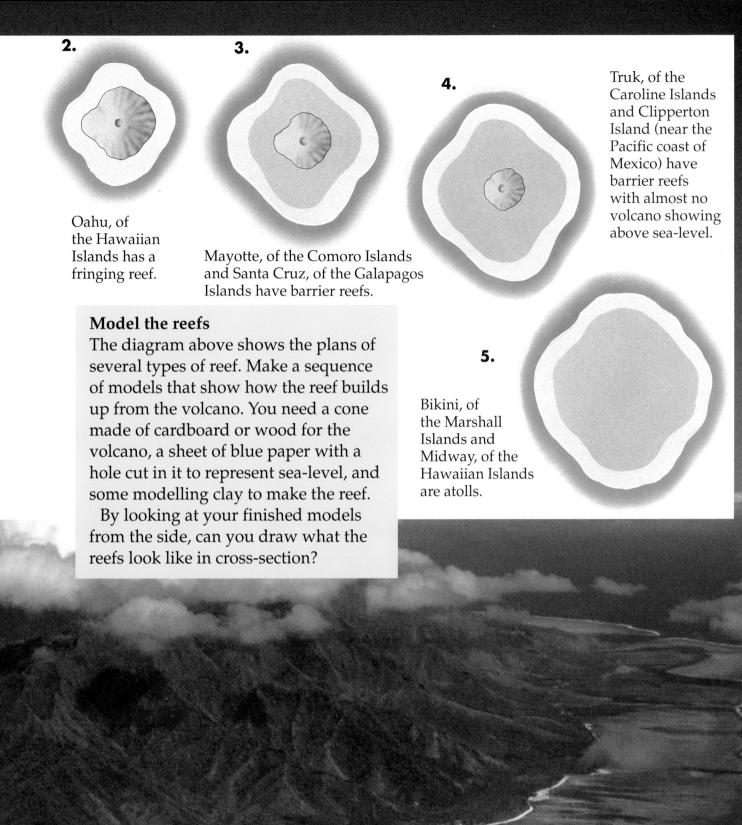

2.

Oahu, of the Hawaiian Islands has a fringing reef.

3.

Mayotte, of the Comoro Islands and Santa Cruz, of the Galapagos Islands have barrier reefs.

4.

Truk, of the Caroline Islands and Clipperton Island (near the Pacific coast of Mexico) have barrier reefs with almost no volcano showing above sea-level.

5.

Bikini, of the Marshall Islands and Midway, of the Hawaiian Islands are atolls.

Model the reefs
The diagram above shows the plans of several types of reef. Make a sequence of models that show how the reef builds up from the volcano. You need a cone made of cardboard or wood for the volcano, a sheet of blue paper with a hole cut in it to represent sea-level, and some modelling clay to make the reef.

By looking at your finished models from the side, can you draw what the reefs look like in cross-section?

Fringing and barrier reefs

The most common types of reef are those that grow close to the shoreline.

Fringing reefs have no lagoon between the reef and the land, whereas barrier reefs trap a band of calm water between the reef and the land.

A fringing reef is built directly from the shore, only broken where rivers bring in silt from the land which makes it impossible for the corals to thrive. See how the channel in the foreground of this picture leads from the valley between the island mountains.

This barrier reef has a clear band of deep water parallel to the land. Notice that only the outer edge of the reef is above sea-level, where it has been built up from coral sand thrown up by the waves.

These people are snorkelling across the top of a fringing reef. Like all reefs, the surface is irregular, with sandy areas and those where the reef is clear. The breaking waves can just be seen in the background.

Atolls

An atoll is the name of a coral reef that is in the shape of a ring (when seen from above). Inside the atoll is a lagoon, and there may or may not be passes or cays leading from the lagoon to the ocean.

Atolls stand just a few feet above sea-level and the crest of the reef is frequently washed by the waves. As a result little sand or soil forms and there will be only a few plants growing on the crest of the reef.

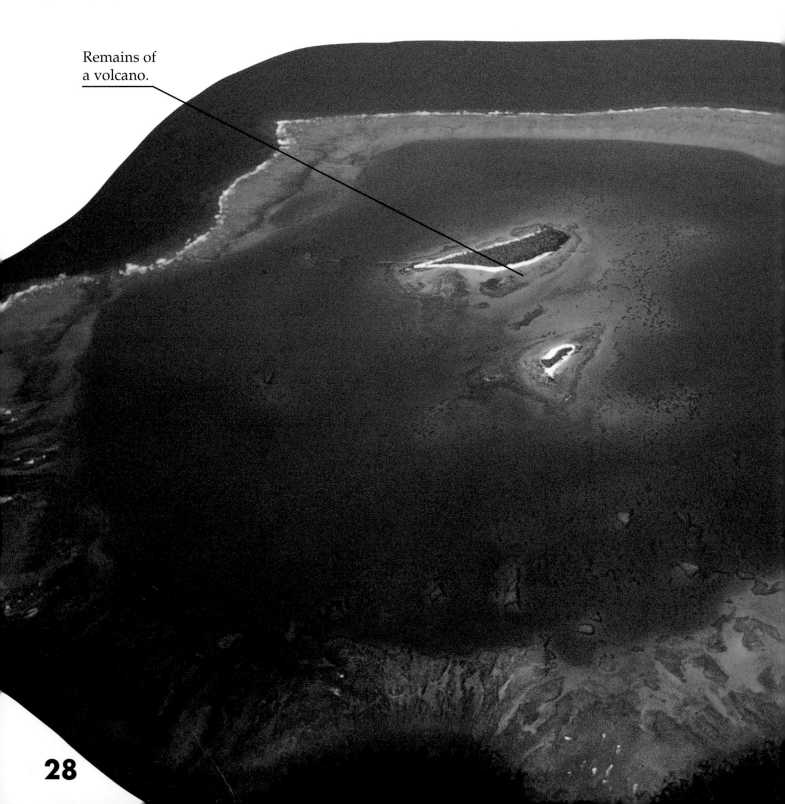

Remains of a volcano.

The large number of small islands show that the atoll is rising from a broad shelf and does not sit in deep water.

Lagoon

A shelf atoll
The picture above shows an atoll among many other types of coral island on a shelf reef. This atoll has a broad ring of coral and a very shallow lagoon.

Low crest with waves breaking.

An island atoll
The atoll in the picture on the left is deep in the Pacific Ocean. In the center you can just make out the remains of the tip of the volcano around which the reef has grown.

Atolls like this have great walls of coral below the surface and the lagoon is quite deep and appears dark blue in color.

The reef crest is flat where sea water rushes back and forth, making it difficult for the coral to grow.

Chapter 3:
Reefs of the world

The Great Barrier Reefs

The Great Barrier Reefs are the largest collection of living reefs in the world. Often grouped together as 'The Great Barrier Reef', it was realised that there are many reefs strung together off the coast and today scientists prefer to use the term Great Barrier Reefs.

The reefs cover a huge area of the shallow sea bed, or continental shelf, off the coast of Queensland, Australia. They extend for 1200 mi along the coast and stretch into the Coral Sea for up to 200 mi.

Why there is such reef variety
When you look at pictures of the Great Barrier Reefs (such as the ones on this page) you will see every shape of reef possible. There are ring reefs (atolls) and ridge reefs (barriers) far out to sea and flat reefs (fringing reefs) close to land.

The Great Barrier Reefs are shelf reefs (see page 20) and they rise from a wide platform. During the Ice Age the sea-levels around the world were much lower than they are today and the early reef was turned into land. On this land rain fell and rivers flowed, carving a new landscape of valleys and hills across the continental shelf.

When the sea-levels rose at the end of the Ice Age, the reef was flooded, the waters warmed again and coral began to grow once more. In some places the land was still flat, and reefs could grow in wide banks as before, but in other places erosion left rings of high land or narrow ridges. It is the coral that grows on these eroded features that makes the 'barriers' and 'atolls' near the edges of the main reef.

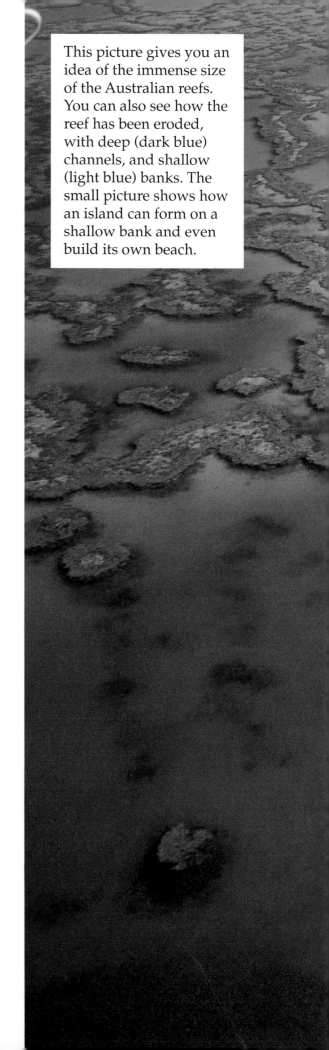

This picture gives you an idea of the immense size of the Australian reefs. You can also see how the reef has been eroded, with deep (dark blue) channels, and shallow (light blue) banks. The small picture shows how an island can form on a shallow bank and even build its own beach.

A satellite image shows the Great Barrier Reefs stretching away from the eastern shore of the continent of Australia.

Islands have developed on top of some parts of the reefs that have grown closest to the surface. The diagrams on page 23 show what the reef looks like in cross-section.

31

The Red Sea

The Red Sea is one of the world's youngest seas. It is surrounded by mountains that are still rising, and its floor is growing as the shores of the sea are slowly pulled apart by forces deep within the Earth.

The warm waters of the Red Sea are ideal for reef life, but the steep sides of this 'new' sea make sure reefs do not develop as wide shelves like the Great Barrier Reefs of Australia. Instead they form a narrow band, clinging to the coast for over 1500 mi.

The narrow shelf of the Red Sea reef shows clearly in the picture on the left, which was taken off the Egyptian coast. The sailing ship is able to anchor close to the shore, yet it is in the deep, clear, dark blue waters beyond the reef. The reef shows as the lighter colored water and the exposed area on the left of the picture.

The Reef plunges sharply just off the coast, giving an almost vertical edge to the shelf. Here, below the influence of the waves, you can see the corals stretching out into the sea to gain the light and nutrients. However, without a sea floor to build on they can only spread very slowly.

The large fish under a table coral is a grouper, the smaller fish are a shoal of Anthias.

Ancient reefs

Reefs are major rock builders and one of the most important contributors to limestone rock. Such reef builders have been around for over 600 million years, and they can be found in many countries of the world, giving a good opportunity to look at some coral of the past.

From polyp to massive cliff
Reef-building animals and plants are each tiny, but in their millions they can make huge layers of rock that will withstand the weather. Many reefs of the past now stand in the landscape as massive cliffs, far tougher that the other rocks around.

On the left is a piece of modern coral, and below is a piece of limestone rock made from coral. You can see how the main stems of coral show in the rock, but that the spaces in between have become filled with muddy lime. This was produced as the coral was buried at the end of its life.

Many limestones that were once coral reefs have been attacked by water and the result has been to create a network of passages and caverns. This picture shows the inside of a limestone cave where the skeletons of coral polyps have been dissolved and then reformed as **stalactites** and **stalagmites**.

For more information on the landshapes of caves see the book Cave *in the Landshapes set.*

Reef limestones are attractive building materials and many buildings have been made with them. When the surfaces are polished the corals show very clearly.

This picture shows the top of an area of limestone. The limestone rock contains many hairline cracks which have been attacked by rain and running water, producing wide gaps.

35

New words

continental shelf
the edge of a continent that happens
to be below sea-level. Most continental
shelves are broad and slope only gently.
Continental shelves in the tropics are
common along the Red Sea, East African,
East Asian, Caribbean, Brazilian and
Australian coasts. They are rare along
the west coast of America because this
coast has many deep ocean trenches

ecosystem
the pattern of living creatures that exist
side by side in an area and which depend
on each other. Often life in an ecosystem
is made into a food chain, with plants
giving food to grazing animals, which in
turn are food for hunting animals. When
the animals die their remains rot and
become food for the plants again

Ice Age
the time, beginning about a million years
ago, when the Earth became much cooler
and ice spread over the northern continents
and out from Antarctica. As more and
more ice formed on land, so rivers no
longer returned rainfall to the seas, and
ocean levels began to fall. When the Ice
Age was at its most severe the sea level
was about 240 ft lower than it is today

limestone
the rock that is formed entirely of the
limey remains of sea creatures

plankton
the smallest forms of plant and animal
in the sea. They are carried along by the
ocean currents, using minerals in the sea
water to build their cells. They are an
important food source

polyp
the soft-bodied anemone-like organism
of a coral. Polyps start life as tiny free-
swimming larvae which find a suitable
vacant spot to attach themselves, then
produce a limestone skeleton. After a
while they divide, creating two polyps
where once there was just one. Within a
short time further dividing has produced
a large group, or colony, of coral polyps,
each making their own skeletons

shelf reef
a reef that forms on the broad, shallow
waters of a continental shelf. Many small
coral islands may rise from the reef and
be dotted over the sea. This pattern is
different from island reefs, which usually
form in long lines across an ocean

stalactite
a long cone-shaped formation that
is found hanging from the roof of a
limestone cave. It is produced as lime-
rich water slowly drips from the roof,
leaving behind the lime as a kind of scale

stalagmite
a long cone-shaped formation that
grows upwards from the floor of a
cave. It forms below a point where water
drips from the cave roof. Stalagmites are
much broader and more massive than
most stalactites

swash
the name given to the part of the
breaking wave that rushes up the
beach. The foaming swash is very
powerful and can easily damage reefs,
but it also carries broken reef material
across towards inner lagoons

Index

algae   4, 9, 14
Anthias   33
atoll   10, 22, 24, 25, 28, 30
Australia   4

barrier reef   10, 22, 24, 25, 26, 27
barriers   30
beach   20
Bermuda   2
Bora Bora   3
butterfly fish   14

Caribbean Sea   2
cave   35
cay   20, 22
channel   20
Charles Darwin   24
Clipperton Island   25
continental shelf   11, 36
coral   4, 9, 17
Coral Sea   30
crest   10
crust   20
cycle   23

desert island   20

ecosystem   14, 36
Egypt   32

flat reefs   30
Florida Keys   2
fringing
reef   10, 22, 24, 26, 30

Great Barrier Reefs   3, 4, 11, 30
grouper   33

Ice Age   21, 23, 30, 36
island   11
island atoll   29

Kenya   2

lagoon   10, 21, 29
light   17
limestone   4, 9, 12, 34, 36

Maldives   3
Mayotte, Comoros Island   25
Melanesia   3
Micronesia   3
Moorea   24

nutrient   4, 17

Oahu, Hawaiian Islands   25
ocean current   16
ocean islands   22
Pacific Ocean   24
parrot fish   15
pass   20
plankton   13, 36
platform   30
Polynesia   3
polyp   13, 14, 36
pools   18

Queensland   30

Red Sea   2, 32
reef animals   14
reef flat   18, 20
reef front   16
ring reefs   30
river   26
rock   34

sandy beach   20
Santa Cruz, Galapagos Islands   25
shelf   32
shelf atoll   29
shelf reef   22, 30, 36
silt   21, 26
snorkelling   17, 27
sponge   15
Stag's Horn   17
stalactite   35, 36
stalagmite   35, 36
starfish   15
stony coral   19
storm waves   20
surf   16
swash   16, 36

Tahiti, Society Islands   3, 25
Tanzania   2
tentacle   13
tide   18

volcanic islands   24, 25, 28

wave   16